MARKETING FOR SCIENTISTS:

EXPAND YOUR NETWORK BY INCREASING YOUR PERSONAL BRAND

C.J. LOCKE

Copyright 2020
C.J. Locke

All rights reserved

SUMMARY

Introduction

Marketing 101

4 p and marketing mix

Positioning

Personal brand

Go online!

Paid advertising online

Stay offline

It's time to act

Conclusion

INTRODUCTION

Whether you are a professional, an entrepreneur, an employee, a consultant, or anything else, it is very important for you to use marketing and communication to get more customers and to keep them over time.

This book is designed especially for you, but has no magic recipes, sorry, better to warn you first.

However, I can guarantee you that by applying these strategies, with a little luck and a lot of hard work, you will be able to see results.

There are no excuses, there are no different sectors, marketing is always the same and works in every sector, whoever tells you the contract is a scammer.

Of course, this book is meant for a professional like you, not for a multinational full of money!

The marketing principles are the same, however, they are simply applied and translated here to be adopted by you from tomorrow.

So let's get started!

MARKETING 101

Marketing is very dynamic.

This includes a variety of topics, strategies and tactics.

Therefore, it can be difficult to develop a fundamental understanding of how it works. Training trade can take years of hard work and honing the skills, often in a handful of specific fields (such as policy, copywriting, or analytics).

Nevertheless, like many aspects, potential success starts with solid foundations.

And if you're trying to learn, you've come to the right place.

How is advertising? How is marketing?

Advertising is the easiest method to achieve effective market activity. This spans the entire spectrum of tactics and political organizations, motivates the target audience to purchase products and services in the sector.

4 P AND MARKETING MIX

Marketing relies on four issues: commodity, size, advertising, and place, for all its nuances. But these are the definition all around and they are concepts which never alter. Strategies and canals shift.

Some of these templates are generalized to 7 P or to a particular variety. But these four should be enough for your needs to develop an understanding of the workings of marketing.

Product

That's what a company sells, be it a physical product or a product (for example, consultation, a subscription or something). The following should be determined from a marketing perspective:

- how many different product variations or lines should be sold? An automaker could, for example, strategize which categories of vehicles to build (e.g. family cars, SUV's, crossovers, or pick-up trucks).
- How are they to be packaged or presented? To give another example, should a company produce floor replacement car mattresses in a box? A bag? A bag? Anything else? What else?
- How's it going to be served? This could include guarantees, returns handling and so on.

Marketers may even be involved in determining how and which features products are designed.

Price

It's only "how many things are costing," right?

Well, sure, of course. But it's more than that.

If it is all about marketing that drives profit, then prices must be set at a market level that supports them.

Here are some marketing considerations with prices: what is the price of a product per unit? This requires some market analysis and competitive research to determine what is a fair price and what people are prepared to pay for the product.

How are discounts to be applied and timed? Should the product be sold at some times in the year?

Is it meaningful to give customers payment options? A car dealer could offer financing options instead of expecting people to pay the full price in advance.

Promotion

Does it even exist if a product is launched but no one cares?

Well, yes, technically it does, but if nobody buys it it, it only takes up space. Once there is a product, it has to be promoted so that people know that it exists.

Which channels will the product promote? This includes online and offline.

Where is it going to be promoted? Online? Online? Offline? Offline? In shops? In the case of events?

What message should be sent? What copy and verbiage will tell the public about the product and encourage them to buy it?

Place

The right product must be in the right place to be found and purchased by people.

Where does the product go? Online? Online? Offline? Offline?

Will certain places receive the product? For example, you might not be distributing as much to Florida if you are selling cold weather clothing and in Minnesota.

POSITIONING

Product marketing is the method of putting the product in the eyes of your consumers. Brand positioning is the technique used to differentiate the company from the others, more than a slogan or a trendy logo.

Who separates you from the competition? Successful companies such as Coca-Cola and Band-Aid all share one important thing: a strong brand. In reality, their brand names have become generic conditions in their market for all similar products. Would you ask for a bandage or a band-aid if you cut yourself?

A strong brand should be a priority for all effective companies, and there is evidence. Brands consistently shown a 23 percent increase in average sales.

B2B licensed businesses achieve a higher EBIT margin than others. And if that's not sufficient, successful branding offers benefits, like increased customer loyalty, a better picture and a related identity.

There is only one safe way to build a strong brand: brand placement. This is how you can position a brand successfully on your market in 2020.

According to The Branding Journal, the degree to which an efficient brand positioning is considered favorable, specific and trustworthy in the minds of consumers can be represented.

Years ago, a soda company decided to do something different and place itself as unique. Today, Coca-Cola is consumed every day as a household product by millions around the world—and it is set in our minds to be the beverage gold standard.

Many sales representatives can use positioning techniques to differentiate themselves from the competition.

Establish an emotional connection with prospects and consumers—Before going to the hard seller, interacting with your prospectives at a human level builds trust and allows your prospects to view your brand more positively. For example, at the start of the sales process, you will take time to learn about your customers and which issues the company wants to solve.

Reinforce the differentiating characteristics of your brand— The distinguishing properties of your business should be easy to understand and relate to with a strong brand role. Make sure you understand what makes your brand unique during the sales process.

Creating value — As a sales representative, the main aim should be to help the client solve a problem or resolve an obstacle. Ideally, the product of your business is part of the solution. You should always look to solve problems for your customers throughout the sales process.

Embody the logo of your client— If you work for a customer, you are your most important ambassador. When you work with prospects, ensure that they have an experience which reflects your company's core values and matches the brand of your company. For example, you should include this language in your sales discussions if your company adopts a light, enjoyable approach to branding. It would not be honest for your business to have an overly serious or static tone.

6 steps to build a strategy for your own brand positioning.

The development of your own brand marketing plan requires plunging into the nuances of your company and figuring out what you are doing differently than anyone else. These six steps allow you to build a brand positioning strategy specific to your business.

Phase 1: Determine the location of your current brand.

Were you actually selling your product or service as another commodity on the market, or is it a distinctive thing? Your present placement of your company provides you with important insights into the next location. To order to evaluate the situation better, you need to consider your current position.

Begin by identifying and describing your target customer. Identify your mission, your principles and what separates you from the rest of the market. Eventually, store your promise of value and your current brand identity and speech for your company.

"We all like ties with companies that looks and look real to us. Instead of creating a complex lingo that nobody is able to understand, only talk to people. Begin studying who the (ideal and existing) target is and use their language," says Matylda.

Phase 2: Competition.

It is necessary to evaluate the competition after evaluating yourself by performing a competitiveness review. Why? Why? You must see who you are against to carry out work for rivals. It work lets you determine what you can do best to make a difference in your plan.

There are different methods for evaluating the competition, among others:

- Market research: remind the sales team which competitors come up during the sales process, or use a market keyword to quickly search and see which businesses are identified.
- Using customer feedback: ask the consumers which firms or items they have contemplated before picking yours.

- Using social media: Quora provides a platform for users to inquire about products and services. Look through these sites to locate players in your market.

Phase 3: Conduct research with competitors.

Once you have decided who your rivals are, it is time for thorough research with competitors. You will have to analyze how your competition places its brand to compete. The analysis should include: what products or services the rivals provide. What strengths and weaknesses do they use successfully? Which role will they have on the current market stage 4? Identify which makes your company special. What is your marketing strategies?

Phase 4: Uniqueness.

The creation of a unique brand is all about understanding what makes you different and what works best for your business. Chmielewska says, "Begin by identifying what' successful' actually implies for your brand, and then build on its identity." You would undoubtedly be beginning to see trends after conducting research by competitors. You will begin to see some companies with the same strengths and weaknesses. If you equate your product or service with yours, your success may be one of their limitations.

That's what makes the company special, and it's the perfect place to put the products on the market. Take note of your unique offerings and immerse yourself in it to identify what you are doing better than anyone else.

Phase 5: Build your declaration of placement.

It is time to take what you have learned to make a statement on brand placement. "A Positioning Statement (Positioning Statement) is a one-or two-state Declaration that expresses the unique value of your brand for your clients in comparison to your major competitors," says The Cult Branding Agency.

What is your type of product or service?

What is your product or service's biggest benefit?

What is confirmation of this benefit?

A clear yet persuasive positioning argument can be created from there. For examples, check out Amazon's declaration of positioning: "Our mission is to become the world's customer-centered company; to create a place where people can search and appreciate anything that they can buy online." We market for everybody a wide range of products, which is also their best advantage. And the evidence? Everything's online.

Phase 6: Does your declaration of position work?

It is just the beginning to take time to position your brand to appeal to a certain customer. When your positioning statement has been made, it is high time that your customers check, try, and gather feedback about your positioning's target.

As Ryan Robinson of Close.io notes, "To place the company in a position to cater to a particular market, customer or audience is just a small part of the battle. It is important to test, explore, and constantly collect (real) input on whether or not your placement has the desired impact from your target customers. We also expanded our roles by constantly asking (and listening) potential consumers ' input as they enter, and our advertising and distribution are clearly an important commodity for our company.

PERSONAL BRAND

How you promote yourself is your personal brand. You want the world to see you through the unique combination of knowledge, expertise and temperament. It's the essence of your life and how it represents your actions, your speeches and speeches and your attitudes.

You use your own identity to differentiate yourself from others. Great, you can connect your personal branding with your business in ways that no corporate branding will succeed.

Your personal brand is professionally the image people see of you. It can be a combination of how they look at you in real life, how you are portrayed in the media and the impression that people get online information about them.

You can either neglect and organically, potentially chaotically, grow your personal brand outside of your influence or help sculpt your personal brand to represent you as the individual you want.

Your personal brand was just your business card during the pre-internet days. Without you or someone who clearly represented you as the promotional picture, few people would have known about you. Within today's highly public culture, where every single move on social media is discussed at length, you're much less discreet.

Your personal brand can be of strategic value to you. This is how you introduce yourself to current and potential customers. It offers you the opportunity to make sure that people see you as you like, rather than in an arbitrary, possibly detrimental manner.

This gives you the chance to highlight your talents and interests. It helps people believe they know you better, and people have much greater confidence in those they feel they know.

This is especially evident at the time of election. While many people are looking for the opinions of politicians on issues of concern to them, the method is less important for other citizens. Rather, they opt for a label they accept. It is no wonder because, regardless of political ideology or persuasion, politicians with strong personal identities excel in elections. Regardless, for example, of your view on Donald Trump, you can not argue that he built a strong personal brand that inspired a large number of people to vote for him.

If you want to be considered influenceful, you need to create a strong personal brand. Your personal brand allows you to stand out. You can use your personal brand to show your skills and knowledge in your fields of expertise.

Your personal brand is in many respects what makes you special. Your own name allows you to differentiate yourself from thousands of others like you.

Millennials, in fact, have a shortage of attention. 84 percent of millennials don't like the advertisements or products that make them. Nevertheless, they are ready to believe individuals they actually "real"—they also dislike business people behind labels.

It required a significant analysis of how businesses sell themselves. In reality, this is one of the explanations why influencer marketing in recent years has been so effective.

There has been a push to personalize the main people in a business. This is obviously easy for a small business—a sole trader and his business differ little. With large companies, it can be tougher. But some people are good at handling it. Steve Jobs used identity messaging well before the term was even widely recognized as Apple's name. Likewise, the personal brand of Elon Musk is probably better known than the company brand of Tesla.

It is fair for any business owner or manager first to connect personally with potential customers before he or she attempts to deliver the message to the client.

To build a personal brand, you have to analyze and look at yourself thoroughly. When you accept yourself, it benefits—something that surprisingly few people do. It is incredibly difficult for most people to identify themselves, although often they find it easier to clarify how they want to be.

To order to improve your business performance, the main goal is to make sure you recognize who your target customers are. You want your personal branding to suit your specific customer.

This is not recent. This is not fresh. This has been accomplished by industry people for many years, long before the internet. Think of Hugh Hefner's image for virtually his entire adult life. Perhaps he never learned of personal branding, but he created the Playboy Empire's logo and enjoyed the lifestyle that his magazine readers envied. But, if he headed a more radical organization, serving more politically right clients, he couldn't live his life-style.

In the end you want to build a reputation as someone who thinks about the kind of people who make your future and existing customers.

A key part of your personal branding is to insure that you are directly faced with the same challenges and issues as your target market. The only distinction is that you can prove that some of these issues—which you are willing to share with others—have been solved.

You don't want to pose like a generic used car salesman, even if you're in the car industry. Cynical customers perceive corporations as offering anything. The goal of personal branding is to step back and focus revenue.

Nowadays, the value of social media can not be overlooked. Part of your personal branding should be social media accounts on all social networks in which the target spends time. And you would like to embrace invitations from friends easily—you won't want to lock your accounts in secret. If you want to actually run a Facebook account for your family and friends, you might suggest having a separate account with a different name.

You want to be associated with your personal branding. You want to stay true to your brand at all times.

This is not just a matter for you to continually behave, although this is necessary, in a way that matches your image. It also contains elements that represent more corporate branding. You want to choose a similar color scheme and fonts. And everywhere you want to use these colors so fonts. These should be the same on your profile, on your social accounts, on business cards and on everything you have made. When you reflect a brand, it should suit your business ' colors and fonts.

Here you want to look at all you publicly use. You even need to look at such things as e-mail footers, paperwork, accounts and invoices and any brochures that you distribute. You may also go so far as to buy a car that matches your favorite color scheme, and use bright ties or other accessories for any public event you attend.

Don't forget the world offline. The personal brand will include all the locations the consumers will visit. Old-style business cards are as important as your social pages to your branding. Just remember to keep your subject consistent.

Carry business cards wherever you go with you. This is just as relevant as it was 40 years ago. Take a card when you can interact with a new person that fits your target audience.

Ideally, you want your personal branding to include your entire public life. This even includes how you look in public whenever you are. I could not imagine Donald Trump wandering in old jeans and a tea-shirt along the streets, anything more than I'd think a rock star would go shopping in an elegant suit. You want your entire public image to match your personal brand.

A secret underlying consistency of successful personal branding. You are trying to establish an official image of "you," so you have to act accordingly.

This is also why the audience will react very differently to the news about the same kind of conduct between two individuals. An act that would surprise them if committed by the first person could be fully consistent with the standards of the second person.

Picture, for starters, the reaction to reports that a certain official had taken drugs. In most situations it is the death knell in your political career—it is completely incompatible with your personal branding and the aspirations of the electorate. On the other side, say you saw a heavy metal rock star take drugs. It wouldn't be true. It would very probably match the standards and could even boost his popularity among his core audience.

The typical person who represents a small business typically has no such extreme views on its personal brand. But you still have to act in ways that correspond to your perceived image. If you are worried about your clients, it is vital that you respond to their concerns and try to fix any issues they may have with your company.

But above all, you want a fairly genuine personal brand. Of example, it may overlook some of your own warts and you may have to tweak them a little to match your target audience preferences. It is not difficult to find a fake person who lives a life lie.

When you are not there, you want your personal brand to suit what people say about you. You risk a significant lack of credibility because people don't believe your personal brand.

You might already know the direction you want to go in your life. In fact, you may be well experienced already and just want to improve your existing personal brand. If that fits you, then please jump to this article at a later stage. But it will still be a worthwhile exercise to revisit your company dream in order to make sure you are headed in the right direction.

You first have to determine what your personal brand feels like before you undertake some workable projects to develop your personal brand. What's your vision?

You'll want to talk about how you want people to look at you. What kind of image do you want to portray?

Just as companies produce vision statements explaining what they want to do, a personal vision statement should be made.

"A simple personal vision is an alignment of your talents, desires, attitudes, beliefs, goals, abilities / experiences, families of birth and the stage of adult growth," the Highlands Company says. Today, it seems huge, but you can really encapsulate everything that is important to you in two to three words.

Of example, the emphasis here is on the dream of the company, so the concentration is on how you want the outside world to see you.

When you build a personal brand, you should look at your interests, beliefs, characteristics and abilities as supply considerations. If you know these, you can probably design a brand that underlines all the good things about you and your strengths. It would show all you have to give to men.

Nonetheless, you are unable to stress everything that you do. In reality, a lot of things that you succeed in are of little value to your personal brand. You might be an outstanding knitter, but today there is still little market for hand-knitted clothes. You may have an extensive collection of English comics and you can clearly discern a Lion from a Tiger from a Beano-however, you certainly can not make money of that experience.

The second area to take into account is demand factors. To succeed, you must emphasize your ability to offer something that people care about. If nobody is involved in what you have to say, your personal brand is meaningless.

Just as a company needs to identify its market, a person creates a brand. You have to think why you do this. Who do you try to impress? There is little point in trying to make someone bland and spectacular.

Go back to your dream. Go back to your dreams. If you want to fulfill that dreams, what are your goals? What moves do you take to achieve your objectives, and who do you convince to make such ambitions a reality?

Ideally, you want to focus on a field where you can demonstrate abilities. To do this, though, a clear target market must be present that cares about the subject. You want to be the one in the niche—the one to whom people have confidence.

The specialty may sometimes seem rather common. While that indicates there is an immense target audience, it also implies you have tough competition. In that case, you could consider reducing your focus while ensuring a large enough target audience still exists to support your expertise.

For starters, you might have figured out that you have experience as a writer. Most people still read books in this internet-dominated era. There are, nevertheless, other novelists too. Most of the best authors focus on one genre. Perhaps Stephen King might compose popular books in many different genres, but he found it easier to concentrate only on building a reputation as a horror writer.

Once you've found your niche, you must determine how different you are from anyone else in your area fighting for the top position. If you are perfectly capable of doing and can provide your target audience with real assistance, what is your difference?

In some cases this refers to the apps you have explored before. What's what makes you distinct? There is little sense in battling your own temperament. It is much better to work with your characteristics.

For starters, Michael Hyatt and Mark Manson are two prominent blogs in the self-improvement niche. They support people who seek ways to improve their way of life and may build leadership skills.

Michael Hyatt is a committed, spiritual leader and Christian. Although he doesn't impose his religious opinions on his family, he also doesn't shy away from them. His posts tend to make moral choices and adopt a Biblical ethic. Living Forward: A Proven Plan to Stop the Drift and Get Life You Want is a typical Michael Hyatt book title.

Mark Manson frequently gives self-help and explains how your life can be changed. He finds himself much more worldly. He's thinking clearly, clear and correct. "I write about big ideas and give life advice that doesn't suck, to quote on his web site's homepage. Most people say that I'm a dick. Some claim that I saved their lives. His popular Kindle title is the Subtle Art of Not Giving a F*ck, the novel name you could never believe Michael Hyatt was using. Learn and evaluate for yourself.

The two people operate in a similar environment but adopt their individual characteristics and establish a clear distinction.

GO ONLINE!

Virtually everyone with a strong personal identity is commonly known because they have a substantial, reliable online presence.

Website/Blog

You want your own page. You want your own. Although social websites like a Facebook site are not yours at the end of the day. Facebook just needs to change or delete their guidelines and the account is finished.

Google controls one of the tools of a successful online company. To do this, you need your own domain, where you choose to use your own name. To conquer Google, you have to produce content consistently under your brand.

Nevertheless, as you meet your personal branding goals, you will make sure that your platform has the kinds of topics that your target audience is involved in. Of example, a personal journal that shows you what you've done today and what you eat can be tracted in Google to find your name. But the type of people you approach will not be of concern.

Of starters, if you were to be a leadership specialist, you would guarantee that your website provides valuable content as a leader. This was achieved well by the previously mentioned Michael Hyatt. The platform attracts people searching for leadership advice and assistance. It helps to associate the word,' Michael Hyatt' with the subject,' leadership.' But a new software update in Google ensures that you can no longer control the topic quest on your own platform. For each quest, Google recommends choosing just one page from a specified web domain. You also want to see the results of the search engines displaying a combination of the best work on your specific blog / webpage, combined with other related web page titles.

A number of services offer free blogs for anyone who needs one. The platforms more common are WordPress.com and Blogger, while Google is searching for a "online blog" that includes over 145 million pages— definitely it fits all your insane criteria and needs. I choose

WordPress.com, primarily because I recognize the industry standard WordPress program used on other websites I am writing on. WordPress.com features built-in spam security, a variety of pleasant topics, a word processor-like text entry feature (no code necessary), good statistical information (to see how much publicity the website gets), and much more. (Note: I don't work or like anything for them).

Your site is constrained only by your passions— what must you say? Think about the intent of your blog-do you want your everyday life to be registered, or a way to keep your family and friends up to date as you travel? Or maybe you want a place to encourage and keep in contact with your followers, your blogging, music or artworks? Perhaps you want to have a connection with your clients— and where can they interact with you? Maybe you want to share with the world your understanding of politics, self-employment, football coaching, high school or fishing?

Nonetheless, you should learn a few details about writing a blog. While there are descriptions of every possible style of writing on blogs across the internet, there are a few characteristics blogging seems to share: it is short: on-screen reading is not as fun as reading on paper, so that people tend to shy away. Two thousand words is long for a blog post; 1,000 words are a pretty good goal; short 300 words or so are completely acceptable.

Paragraphs are shorter: because you have to navigate a lot on a page, paragraphs are usually shorter so that a whole idea will fit into the browser's frame.

Important things are emphasized: internet viewers tend to skip bits, so site writers frequently position key points in a bold manner so that their followers can choose the most important things quickly.

Bulleted lists are common: Bulleted lists are another way to accommodate skimmers, and allows all the most important points visible quickly.

This contains links to other websites: posting aims to use the ability to link to your job, whether to provide explanations or start discussions somewhere (e.g. a connection to a description of a complicated term or idea of a Wikipedia) (e.g. a link to a new post on a forum to which you reply).

It's talkative in tone: blog writing tends to be slightly more personal than most writings. What readers seem to react to is the unique voice of the authors, his individuality conveyed through writing. Which ensures that you can use "I" and "you," use language and even curse if it is appropriate for the intent of your web.

Of note, all these "laws" have been repeatedly broken.

Millions of blogs are on the Web, but only a small fraction of them are involved. There are many explanations why bloggers "die "— they don't have anything to say, they're distracted, or worse, they like they talk to themselves. Here are a few suggestions for holding your blog up to reading: Making a plan for posting: Before you launch your site, make sure that you write once a week, or every two weeks and delay your calendar for some time. Start at a low posting rate— if you assume you have enough time to post more, it'll be a pleasant surprise for your readers (unlike reading frustration when you publish every day every two weeks).

Sit down with a journal and write down a hundred (or 50 or 200 or whatever) things you might think about. Set up the bar big. Better yet, build 100 titles of upcoming blogs. You can even take another step to compose short sketches how the post can appear, get out your journal and ff your list if you are lost for something to write about.

Write posts in advance: create a list of 3 or 4 posts. It offers you a coat if you are lost on a topic and if you are on vacation, you can also use the postal schedule option of serious blogs (WordPress.com and Blogger all allow) to plan articles in the future.

Say your address to people: Let people know you've got a blog. Place the URL on your business cards, attach it to your e-mail signatures, add it to your social networking sites profile, add it to your blogs, and so on.

Post on forums for other people: Be an active part of the writing world. Users will see the feedback and clock the button to know more.

However, in your area of interest you can make friends.

Connect to other blogs: when other bloggers see that you are connected to them, they're going to search out you— and connect back to give them popularity.

Write a guest post: Many blogs will publish guest posts by other bloggers to help them get attention. See your favorite bloggers and see if they have contributed material — if you can't find it, contact the author to inquire.

Write great content: I saved the last thing that is most important. When you write badly, or if it is dull, no one will visit your article. No one is going to publish your guest posts, and nobody is going to connect to you. You will not be inspired to compose because you will believe that no one is reading you and you are not satisfied with the quality of your writing. You don't have to be Hemingway, but you have an authentic, entertaining voice to create.

Graphics

Each site has an eye-catching logo on every page, which is often reflected in the favicon of the site. They typically also repeated their color scheme and fonts on their social pages. They settled on a clear set of fonts, and each of them implements a color palette chosen by the artist.

While some chose a simple, minimalist appearance, others took a more luxurious and detailed appearance. The layout of each website primarily represents the interests of the target audience that regularly visit.

If you were to follow one of these people on the internet, you would notice that they have always looked everywhere they have an online footprint. You will even note that they generally use both their social sites and their forums in the same profile.

The part looks like a big part of building a personal brand. It is like your house to your blogs and social websites. You invite your target audience, and you want to look like you're there.

You don't want to skimp on a custom logo. It must not be sophisticated, but it must represent you. One designer can be found to make one for you, perhaps on a site like Fiverr. You might set up a project like 99designs at a crowdsourcing platform.

Guest blogging

You can then consider posting visitors on other similar sites once you have built your own platform and produced quality content accessible to everyone. If your guest blog posts are popular, they do appear in searches for your name—and ideally in searches for the right keyword words in the topic. The idea behind guest blogging is that the audience in your niche is "borrowing" a popular blog.

Effectively, a blog welcoming your invitation for a guest post advises you, "if you are able to write my audience article, I'm willing to let you put a small ad at the bottom of your message." Yes, your ad is going to be a link in your guest post office to the correct page of your own website.

However, when hunting for guest blogging opportunities, you must be strategic. There is little point in writing a blog that does not reach your own target audience. You should find just posting visitors for your community on the correct forum.

Social Media

William Arruda, "The Personal Branding Expert," and founder of "Ditch, Try, Do: 3D Personal Branding for Executives," suggests, "if you want to be effective, you need to think of yourself as a personal brand." Like a company brand that reflects a certain corporate entity, an individual brand is an independent example of "you." But not everyone can carry this off effectively. If you have struggled to market yourself, social media is where you should continue.

Social media is a powerful tool for creating a personal identity, your image and your business.

It is used by marketers in all fields as a key branding tool.

The amount of benefits it brings to one's company and career ambitions makes personal branding in social media more and more relevant.

But in one day you can't build your personal brand. This calls for constant effort and the right mix of strategies. Here are some useful tips for you to learn how to use social media to market yourself.

Defining what you want to achieve with your personal brand is very important. You must set your branding goal. Would you like to start a new company? And separate the brand / products / services from the current competition? And benefit more by increasing sales?

When you decide what you want to do, you will influence how you want to market yourself on social media platforms.

Let's say, for example, that people recognize their professional skills and increase their chances of finding new jobs in their field. If you want a broader audience to be identified, you can build your Twitter profile by using popular tweets that will draw a significant number of followers. If you specialize in photography, Instagram is your place. If you are an arts and crafts specialist, Pinterest should be your dream choice.

The next big step in social media personal branding is to identify the specialties. This allows you to identify the name.

You must showcase what you stand for and how people should recognize you whether it is your personal or business brand.

Include your passions and skills in all details of your social media profile. It is used as keywords to help your brand appear when people look for such information in the relevant search results.

When you advertise on social media, you want the public to know you and feel better connected. If someone feels connected to a brand or personality, better confidence is more likely. A marketing statement is one way to momentarily, but clearly defines you as a company, bind your audience. In making a marketing message, note that your experience, the ideals you advocate behind, the goals you want to accomplish and the dream you are embodying should be emphasized.

A brand identity lets an organization place itself in culture or in the sector. But it's not only about businesses. Many well-established individuals have their own brand declaration.

The problem now is: will the brand slogan be a representation of you and your ability alone? Not required. Not inherently. A personal brand slogan works best when the interests of the people you are trying to reach are expressed. It should help you relate to your target group.

For starters, Jeff Bullas urges his listeners to "benefit in business and in digital life," and it is crucial to have a clear web presence in social media branding. It's ideal to have several social media channels to reach your target audience. You should have preferably a personal blog and a page for each of Facebook, Instagram and Twitter's main social media accounts. You should also try a YouTube channel if you want to create video content.

Like a corporate branding strategy, your personal social media branding should include a consistent content strategy. You will focus on developing quality content that you want to read and share from time to time, if the aim is to drive traffic to your website.

You will build a calendar with material determine what to publish and when to post.

Social media is also a forum for interaction. And you can't always promote your company. You have to build a reputation to fight this when you brand yourself on social media. Your personal brand should provide you with more individual experience, knowledge and skills. as an individual.

When you are consistent in voice and brand marketing across all social media channels, you will build a better and more impactful personal brand. You want to advertise yourself in a manner that you know how to promote yourself on social media so that people can remember the name quickly. A simple emblem and similar design elements would be an indication of this.

For example, if your personal blog has a red theme, take the same look on your Facebook and Instagram accounts. You will be remembered by your clear profile pictures and the branding used by your personal brand.

The production and distribution of high quality content through social media also boosts the name in Google's eyes. It also promotes better participation because people like to read and share relevant information. You should also periodically post content generated by reputable sources in your field in addition to creating your own content.

It is also very important to ensure that the new information is available on all the social networks and website links. Social messaging means ensuring that your viewer considers you important. Having your profile connections up-to-date would help promote the new work and drive your networks more traffic.

Importing your Gmail addresses and your telephone book lets you communicate and increase the number of people on your social networks.

In recent years, various platforms have enabled users to increase their audience's involvement. When you brand yourself in social media, do not be afraid to use tools such as daily stories to keep your audience engaged in real time brand updates. Data has shown that background features on social media sites expand 15 times faster than newsfeeds with over a billion users now benefiting. Such regular reports have been added on Twitter, Instagram, Snapchat, WhatsApp and YouTube. You will hold the followers up to date on any forum at any time.

If you mark yourself on social media, you can get engaged with your followers by inquiries, debates, responses, replies etc. This can help you establish a strong partnership with your customers and maintain your supporters' allegiance to your company.

Find and join groups that are relevant to your field of expertise. Such groups help you build awareness, develop new concepts, test your skills and build confidence. You will be able to expand the network and build relationships with like-minded people. This is a crucial step in building your personal brand.

Imagine a relationship with an influencer before choosing how to market yourself on social media. The founders in social media are influencers. The increasing popularity of social media makes them the most important medium for brand building and promotion.

Such influencers are people who follow a big fan. They actively interact in social media networks and reach their followers with their creative content.

Working with and linking to main influencers in your market will help you build your brand. Influencers connect you to their followers by creating content on your brand. It offers your professional identity on social media credibility and legitimacy to help you gain the confidence of your target audience.

People trust and follow their suggestions. If you talk about your products on social media networks, influencers become the face of your company. This can help you raise your knowledge of the company and add important information / traffic to your web.

Nevertheless, to accomplish your branding objectives, it is crucial to approach the right influencers.

If your relationships do not contribute to your specialty or area of expertise, your attempts will be in vain. Use a marketing tech influencer such as Grin to identify and link important influencers who can assist you with your personal branding activities on social media.

This takes a great deal of energy to market yourself on social media and you will want to insure that the outcomes of this time and effort represent that. And figure out whether your personal branding strategy works, you should test social media metrics. It helps you understand whether or not you have the right audience and if you are understood. If not, you will strive to diversify your content strategy and increase the effect on consumers on social media.

Build strong ties with influencers in social media so that you can potentially become one of them. Try to create your own fan by exchanging relevant content and common interests, and by writing for them, make your audience feel significant. You can therefore create your own profile on your chosen network and accomplish your personal social media identity.

The answer to the question, "how to brand yourself on social media," begins with you as a leader. You must make an impression in your social circles and communities in order to do so. You will also need to be active in main networks such as Instagram, LinkedIn, Youtube, Google+, Snapchat, etc.

The strongest medium for creating your personal brand is social media. It is safe, comprehensive and has enormous potential.

Social Media Compendium

1. **Facebook** is a popular website for social networking, where users can post feedback and share any information, from images to news. Maintain your interest in your brand by regularly sharing updates. Posting once a day is perfect, but it can be more reasonable at least three days a week.

Guests to like your website is one of the easiest ways to increase your followers. You can also run Facebook ads and target the community according to race, age or venue. In order to build on your Facebook network, it's important to love and comment frequently on quality content.

2. **Twitter** is a strong personal branding tool for founders of startups. Nearly 40% of Twitter users claim that they have made an investment directly from an influencer post. Active every day on Twitter is a good rule of thumb. One to five posts a day will dramatically increase the audience engagement.

Leading crowds is one of the easiest ways to attract men. It is a good idea to monitor the industry at least two individuals a day when you launch your network. Taking part in Twitter talks is another way to increase your number.

3. For any specialist, **LinkedIn** is an utter necessity. Begin by creating a profile that allows others to know who you are. Make sure you create a company profile also if your company does not already have one, so that you can add it to your profile.

LinkedIn provides a little more flexibility than other social media channels in posting rates. The designers will post at least once a week, but not more than five days a week, according to the CoSchedule. Coherence is essential in the development of links, so you should post articles and engage regularly with groups and connections. Choose and adhere to a routine (daily or alternating days).

4. **Instagram** is a multimedia medium, which makes it important to build a personal branding plan to share your know-how through images, videos and text captions. Build an Instagram account that combines your personal with your company.

Founders will try to post once or twice a day on Instagram. One of the best ways to help others to find the material is to use hashtags. It creates new Instagram followers, which could lead to new buyers. Be cautious not to overuse hashtags because your post may be incorrect with spam. Try to stay 5-12 hashtags. Make sure you like and reflect on the material of others to improve your following.

5. **Snapchat** Developers can use Snapchat to create their personal brand as an effective messaging device. Post pictures, photos, and pictures to make a story on your Snapchat story that lasts 24 hours. Content updates can hold the company in your followers ' minds fresh

every day. Edit your Snapchat story after four to five hours or three times a day.

Raise your followers by taking over your account the day, or by adding another page. Promote the split of your supporters and your friends. On other marketing channels you can also advertise your Snapchat account, so that the present public knows where to locate you.

6. **Quora** Executives can use Quora to show their experience in the market, create reputation and boost their personal brand. There is no rule on how often you should post on Quora, but it is important to be active consistently. Try to answer questions, leave comments and vote at least three days a week on the responses of others. Place yourself in one niche so that people know what to expect and want to follow you. Gain visibility by posing insightful questions about strangers.

7. **Reddit** Sometimes known as the front page of the internet, Reddit is a wide community site where people can share information, comment and submit or import articles from others. Collectivities are classified into "subredits" that cover a different subject. Founders are invited to

join the community and make friends in order to help build their own personal brand. Concentrate on being interested in subjects (subredits) connected to your experience by answering questions and giving detailed answers.

Try posting on subreddits a couple times a day. A follow-up is important to begin building a personal brand on Reddit. Add value to each post and make your comment to expand.

8. **Medium** is an online publishing platform for readers to write articles. The Website allows everyone to create content and build their identity as an authority in the field without sharing specifics of how to set up their own profile. Use the medium to publish content related to a brand or enterprise theme. Schedule a regular posting schedule to keep the readers up to date, once a week or once a month.

9. **Pinterest** Start to grow your personal brand on Pinterest by setting up an examination page. Make sure you check your account so that you see the Pinterest logo on all pins on your blog instantly. The primary purposes of your profile are to attract attention to your platform, to

store all original pins from your account and to look authoritative and detailed.

Pin to create your personal brand continuously and through website traffic. Pinning a few times every day or every other day is more successful than pinning mass once a week.

10. The **YouTube** platform enables you to create a personal brand for your own channel and post videos. First of all, make sure you define your personal brand and publish videos based on your interests and profession. Of starters, you may want to share videos focusing on specific marketing areas such as email marketing benefits if you are a marketing expert.

As with many other social media channels, YouTube can deliver the best performance reliably. Post once a week, if possible. If this is not possible, every two weeks you should update.

PAID ADVERTISING ONLINE

Facebook & Instagram advertising

Each time the word Facebook is used, it is implied that the same instructions also apply to Instagram, as the operation is the same and both social networks belong to the same company.

Around 2004, nobody would have assumed that Facebook is what it is now. Facebook is a big part of people's lives with more than 2.2 billion monthly active users and almost 1.5 billion active daily users. This is a phenomenon that will not alter as long as we go through the waves of the news era.

Facebook is perceived to be one of the best online advertisement platforms and digital ad revenue accounted for 51% of US ads.

Facebook's annual advertising revenues in the last decade have steadily increased over the past year because Facebook ads work. In a Kleiner

Perkins study, 78 percent of Americans claimed they found goods on Twitter.

Facebook ads can also work for you if you spend time learning the technical art of producing highly efficient Facebook ads.

As of June 2018, 55.1% of the world's population had internet access, and Facebook has great growth opportunities now and in the future as more users access to internet and enter Facebook. Facebook and the Web as a whole remain the best way to connect with and generate sales from advertising for customers.

It takes little time to create a Facebook campaign and can produce excellent results for your company. It's a clear truth that digital advertising produces a big ROI and with the right campaign and product it can happen very quickly. That's why you see Facebook ads for both small and large local businesses.

Everyone can sign on to Facebook, build a business manager page, set up a payment method and post a campaign in minutes. You can even check the advertisement with ad creation tools, including AdEspresso, to find the lowest possible cost per lead.

We can see that from 0.10 cent by press you can start driving traffic to your website. To get going, you do not need an extraordinarily large budget and it's quite easy just start using Facebook ads compared to running ads on other digital channels such as Youtube.

To attract users directly from your Facebook page, or to build a more nuanced plan to target individual customers and achieve higher outcomes with the business manager, you may raise a single post.

Using this tutorial, you'll be sure how to build a successful Facebook advertisement using business manager.

You can include step-by-step guidance on how to set up your Facebook ad page, including the guidelines on how to work with Facebook Business Manager and Facebook Advertising Manager.

What is the Business Manager of Facebook?

The Business Manager is a device that is intended to handle the Twitter and marketing sites.

Using Business Manager, you can: monitor connections to your Facebook pages and ad accounts—see what people can access, remove or modify permissions from your pages and ad accounts.

Working with agencies—your Business Manager account can also be shared with agencies to help you manage your ad campaigns.

You can use the Business Manager password to combine multiple ad channels and applications.

You can also control and change a different set of positions for Facebook ad accounts in the Business Manager.

Your business manager must manage at least one Facebook page to create a campaign.

To link a Facebook page to your business manager: go to Business Manager Preferences on the left, click on "Accounts-> Sites" in the column of the sites and click on the "Link page" CTA Choose some choices from the three options of the Facebook page: "Demand entry to a profile," "Build a new page" etc.

Before choosing your payment method, you can not post your first advertisement. This pays for your ads to run. You won't have to do this again when you set up your ad account until you have to adjust the payment method, or create a new ad account.

Let's navigate to and set up our ad account settings!

Go to the Business Manager and press on "Add Funds" under "Accounts." Complete all blanks by entering the name, address and other relevant information of your company.

You will also be asked to enter your VAT number if you are a business in the EU. You can choose your billing currency and timezone on Facebook. Be very cautious with this detail, because once you have established your ad account, you will not be able to change it.

You will need to input your payment details after you submit your company information.

Now is the time to add our payment method and build a Facebook page. Click on the settings of your ad account-> Payment Settings. From here, our cc or payment method details can be inserted.

You will, of example, be on the "Campaigns" tab to build a campaign. Tap on the green "+ Create" CTA from there to start a new initiative!

Now you can choose from a handful of campaign targets that match your advertising objectives on Facebook. You might use "Area knowledge," for example, if you want to drive traffic to a physical location. You want to use "Conversions" if you are driving traffic to a page.

It is crucial to choose your target as Facebook uses it to define certain aspects of your campaign, including different ad forms, bidding opportunities and how the advertisement is optimised.

You should always pick the target that best reflects the desired results. When you support a platform, a marketing drive is most likely running. This is the type of campaign we are going to create today.

If the end purpose of your campaign is to provide you with information via a lead form for example, you should instead select the lead generation objective.

The next step is where your adsets or audiences will be created.

When we look back at the last segment, the key marketing manager screen should be recalled with a campaign tab, adsets tab and an ads tab.

The advertisements and adsets include a specific combination of innovative ads and a particular adset with a specific audience and expenditure for the advertisement.

When you develop your following, Facebook will display you your average regular reach and inform you if it is too that, too low or just correct to use the meter. In your campaign, you will track "likes" or desires on Twitter, actions or trends to set up very specific audiences for your campaign.

It has an unrivaled value to set the right Facebook target audience for the final results of your ads, and we will explore later how to construct a custom or lookalike audience.

Most people find it too complicated, clunky and frustrating for Facebook ad managers to use. This is where AdEspresso is performing. AdEespresso is a web-based platform that allows you to create complex ads and break experiments without using Facebook ad manager for any additional fluff or hassle.

There are a number of benefits in using AdEspresso:

It takes less time to set up publicity campaigns with multiple creatives and copies

You can access the features of Ad Manager without things getting complicated.

You can see advanced campaign reports and optimize your ad campaigns.

The advice is to try and experiment, in order to understand how the ads on Fb and Ig work. Then you can decide to use any tool or to outsource this task to some agency.

Google Ads

When you consider spending money on advertising to meet your target audience, it is easier for you to invest it in the right place. Somewhere with more than 246 million guests, 3.5 billion connections a day and an impressive 700% return on investment.

Something like... Google Ads. Google Ads.

Google Advertising was only introduced two years after the world's most popular website: Google.com. The ad network was introduced in October 2000 as Google Adwords, but it was called Google Ads after some rebranding in 2018. Given the expansive Google reach, you probably saw (and probably clicked) Google ads... and your potential customers did.

The guide will teach you what Google Ads is and what you need to know in order to launch Google ads. We will discuss product functionality and show you how to automate your promotions with your advertising to achieve the best impact.

No wonder, the bigger and more oriented the paying advertisements these days, the more views you generate — this contributes to a better chance of reaching new customers. This is why Google Ads is becoming increasingly popular among businesses in all industries.

Google Ads is a paid advertising site, which is a pay-per-click (PPC) marketing system, on which you (the advertiser) pay per click or print (CPM) on an ad.

Google Ads is an effective way to drive qualified or fit clients to your company when seeking products and services such as those you offer. Google Ads helps you to raise traffic on your page, gain more phone calls, and increasing your visits to the shop.

Google ads allow you to create and share timely ads for your target audience (via mobile and desktop). That means you're searching for products and services like your own through Google Search or Google Maps on the Search Engine Results page (SERP) at the moment. This way, when it makes sense for you to reach your target audience.

Note: Portal advertising can also be spread across other platforms such as YouTube, Blogger and the Google Display Network.

Google Ads will also help you analyze and improve these ads over time in order to reach more people, so that your business can meet all the campaign objectives you have paid for.

However, regardless of the size of your organization or finances, you can tailor the advertising to fit your budget. The Google Ads app helps you to remain in your monthly limit and even interrupt or delay your ad spending.

Google is the most commonly used search engine with 3.5 billion search queries a day. Not to mention, Google Ads has been around for almost two decades now, granting it a certain seniority in the world of paid advertising. Google is a tool used by people around the globe to ask questions addressed by a mix of paid ads and organic tests.

Yet Google says marketers earn $8 for every $1 they pay on Google Ads. There are also a few explanations why you'd like to find Google ads.

Do you need a different reason? Your rivals use Google Ads (and may even compete on your company terms). Thousands of companies use Google Ads to advertise their products, ensuring that even if you're organically selected for a term of quest, your search results are pulled behind your rivals.

Google Ads should be part of your paid strategy when using PPCs to advertise your products or services—it's impossible to do so (except perhaps the Facebook Ads, but that's another article).

It is relatively easy (and quick) to set up your paid campaigns on Google, mostly because the site guides you through the setup and offers helpful insights along the way. If you enter the Google Ads site and press "Start Now," a number of steps are taken to start your ads. If your ad copy or photographs are made, it should not take longer than 10 minutes to set up them.

What is less evident are all the additional things you need to do to insure that your advertisements are optimized and simple to monitor. Together, let's protect all. These are the moves that you will follow while your advertisements are checked.

You will probably have Google Analytics set up to track traffic, conversions, goals and any unique metrics on your website. You must also link your Analytics account to Google Ads. Linking these accounts would make it much easier to monitor, interpret and communicate across platforms and programs, because you can display all activities in one location.

Google uses Urchin Tracking Module (UTM) codes to detect behavior related to a particular source. You never encountered them before— it is the portion of a URL that goes after a question mark?"). UTM codes tell you which bidding or advertising led to a conversion to track the most effective parts of your campaign. UTM codes make optimizing your Google Ads easier because you know exactly what works.

However, when configuring your Google Ads, the trick is to add your UTM codes at campaign level to prevent you from manually using every ad URL. If not, you can manually connect them for Google's UTM app.

Conversion tracking informs you just how many clients or routes you bought from your ad campaigns. It is not mandatory to set up, but you'll guess the ROI of your ads without it. Conversion tracking helps you to monitor the website's revenues (or other activities), device downloads and ad calls.

There is something to tell about having all your data where you can store, review and monitor. To order to track contact data and lead movements, you already use the CRM. Integrating Google Ads with your CRM allows you to track the ad campaigns perform for your customers, so that you can continue to target them with specific deals.

Again, the same advice previously given is valid, try! Instead choose what to do.

STAY OFFLINE

Yes, the online world is nice and nice, but people need physical presence, so it is very important to develop your presence in the offline world well.

Below I will tell you about the best way to increase your presence in the tangible and real world of things!

Wear your brand

Create a fashion statement by displaying your staff uniforms or other products with your emblem, slogan or hashtag on it, while you advertise your company. Good design and high-quality materials are important to stand out well.

Let your workers choose from a range of shapes, designs and colors to convey their personalities. You will print extra shirts for sale or for free contributions at special events or promotions. With the use of your hashtag in the logo, you improve the popularity and chance of more use.

How it suits your digital strategy: Through literally having your employees carry your logo, you'll continue to increase the unified identity of your social media by posting photos from your workers or clients at activities that display your swag. Furthermore, more offline spaces will promote your brand name and hashtag.

Your space for local meetups

Act as a partner to a conference or civic organization by providing the space for use at conferences, discussions or gatherings. Most interests or special interest groups find it difficult to find accommodation for their expenditure and group size.

Entry to communities which are in line with your desires or ambitions, and you will probably earn the trust of a few new customers, plus good word-of-mouth connections and some social media posts (especially if you question them in return). It is a perfect opportunity to show how your company also supports your community online.

How it works into your digital strategy: from it you'll get many social media miles, especially if it's a continuous occurrence. Share your social media and newsletter, and share it on your channels.

Host

You may not think of events as offline marketing, but they are a great way to get people to the door. Throw a party or have a special promotion when you launch a new company, release a new flavor or a new local retailer partner. Seminars or knowledge sessions and seasonal events can also be arranged. Events do not need a large amount of space or expenditure to function.

Through holding a pop-up show, reading slam poetry, a modern planting workshop or a group Yoga class, depending on the product and target clients of course, you can become even more innovative. Contact community members and local specialists who may be interested in organizing the event with you for these events. You should also have a network of people you meet who are involved in.

How it works into your digital strategy: fun or educational lifelong experiences that inspire people to post from the experience. These are also capable of digital advertising.

Networking

Profit from your proximity to other organizations by working with them for a one-night or a weekend function. Build a customer appreciation event, scavenger hunt or free day and allow people to visit the destinations for a certain price to which everyone contributes. You will attract many new customers and build relationships with your neighboring companies whilst you are there.

They can also collaborate with organizations who share their values to help customers accomplish the same mission (such as good health or a certain lifestyle), but with another product or service that fulfills that aim.

How it suits the digital strategy: Tell on a Facebook event and retweet to improve exposure with other organizations. You can buy your own ad and advertise it further and, of course, utilize social media or digital signage to increase awareness.

Referral

Print reference coupons and credit cards and offer a free gift or discount to customers who refer other individuals to you (and to those who benefit from this reference). Returning clients should also be compensated, so give them a small bonus for their 10th day, or every time they spend 100 bucks. A little consumer support goes a long way to holding the customers faithful.

How it work into your digital strategy: A paper coupon or rewards card complements your digital coupons or electronic search scheme. If you already have digital displays in your shop, build logos that promote these awards and show them on your screens. You can also advertise the incentive schemes, or even incorporate them into your mobile app for consumers on Twitter and Instagram.

Local influencers

Many of these tactics have been designed to help you build organic word-of-mouth communication through your customers, but another way to make sure you meet as many people as you can in your target audiences is to partner with a credible influential individual or network of influencers who have both credibility and impact in your sector. A respected advisor will sing your devotion to his fans far more easily than to meet certain people individually and get them to take their word for it.

According to HubSpot, word of mouth impacts online conversation about nine times more than likes, shares and retweets, and research shows that thousands of years have more customers that companies confidence.

Just do your research and make sure both methods match together. You need to turn over some creative control, for example, to a famous blogger who publishes their own brand material. On the other hand, if your product or service seems like a weird fit for an influencer, its followers might get irritated. It should be normal for this individual to support your brand.

Guerrilla marketing

The standard definition for guerrilla is a party of rebels who seek, for the sake of a greater cause, to provoke something against a normal faction. Through simply linking the word to ads, you use tactics and promotions of guerrilla marketing that annoy customers in your daily routines. Such distractions are not disruptive, just special, unusual and innovative techniques that open the door to brand contact and retention. The aim here, by drastic approaches, is to inspire something deserving of coverage, to stimulate attention, to get viral and to render circulating so that people spread it by word of mouth or through social media. Guerrilla marketing is the polar opposite of traditional marketing by relying on unorthodox and non-traditional approaches.

The expense of the guerrilla marketing plan for many marketing agencies is relatively low. Such a plan will not break the bank, so that everybody can, including small businesses. Basically, the development of revolutionary material reworked or re-proposed the existing content. You can choose to take several highlights or operate through multiple segments, to then extend material to include various things such as blog posts, memes or photos. The guerrilla marketing requires more than money to invest time and effort in the production of the creative and eye-catching material that the public would feel obliged to share word for word. It encourages loyalty, raises awareness and enhances the brand's revenue.

Guerrilla marketing gives you an opportunity to create a WOW effect and to be unforgettable for your target audience. Guerrilla marketing forces you to churn something unique about which people are going to speak. You quash the same old campaign campaigns and come up with better strategies that give people who experience the approaches a little shock. It is much easier if the viewer has a unique experience with your company. Moreover, with guerrilla marketing, you give your marketing team the opportunity to truly brainstorm and work out ideas that boost their trust and creativity.

IT'S TIME TO ACT

I know that the path I have proposed so far may seem difficult and long to complete, but one step at a time you will see the results, below I propose some small ideas that you can apply immediately after completing the path or even earlier if you prefer.

They are simple and immediately applicable ideas.

Stickers

Easy stickers help the viewer get to know the brand better. It is not a traditional marketing strategy because it is small, but it has the ability to reach a medium size despite its size. The Reddit website used its famous alien image stickers. Their bikes, boxes, sticks, busses, binders, and more captured them like a wild fire. Eventually, sticker gatherings and sticker activities took place. Later, images of stickers and people who stick the alien logo were embedded in all social media. The CEO expressed that they invested less than $500 on these stickers, but the returns were massive. Reddit now has 36 million user accounts and 169 million regular monthly users. Always underestimate the power of the basic adhesive to improve your brand.

Installations

The creators of the 2017 smash horror movie entitled "It" hired red ballons by this guerrilla marketing strategy, binding them to drain grates in Melbourne, Australia. The prank used the audience's association with the iconic red balls as it is related by a character called Stephen King Pennywise, a deranged clown that was prey to teenagers, from his best-selling book also known as, "It." They have created many murals on the mask of Pennywise throughout the region. These strategically positioned murals and ballons built up people's anticipation of the film, and these methods resulted in massive traffic in social media for the film's trailers, leading to the film's success. Think of how you can use these special services for the goods of your business.

Partnership

Partnering with a brand, especially an unexpected brand, will sometimes help promote your goods. Colgate Max Night has collaborated with nearby pizzerias to show this. They offered them customized Colgate boxes for their meal. The client should not forget to brush their teeth before night.

Flash mob

To support your brand, arrange a flash mob system. It involves a group of people executing specific action sets at a designated time and place. You should hire actors or invite leaders of your community to help you build a flash mob that certainly would allow photos of everyone around you. It's nice to raise interest and create online traffic when the public shares their videos online.

Survey

Not only is a fantastic way to get to know your customer base, a survey is also a perfect way to remind consumers of your life. Surveys were number two on our list of ideas on small business marketing.

Find this like illegal email marketing—but this can be achieved in your shop or on the street absolutely comparable. Customers would understand that you have sought their feedback and will feel invested in survey conclusions, which over time will turn into greater loyalty. You may learn something or two about how your company can improve along the way. Win - win.

Milestones

This could be one of our favorite free marketing ideas for small businesses.

Crush your company and figure out if any important information falls in—perhaps you have done business with over 1,000 clients in your field, or perhaps sell your service to a percentage greater than industry standards.

Publish the results online or add a sign above your shop. People respond well to numbers— it is easily digestible. "The community loves us," for example, is "over 2,500 satisfied customers in our city!" Each day. Every day.

Conversely, claim that a university study circles around the business. Use it to add to your company's value by posting it publicly and illustrating key parts.

Infographics

With all these data speaking, it is worth noting that infographics are super powerful marketing tools–and relatively simple and economical to make.

Interactive and quick to grasp infographics are why so people like to post it. Digital resources such as Visme can be used right in your browser— no tech or design skills required.

You can also take a note and pay a professional artist to create a set of infographics for you—a relatively cheap concept for small business marketing.

Online contest

Prizes don't have to be extravagant — a few free products or supportive benefits can be enough to encourage potential customers to participate in winning applications for a few seconds.

Contests are a perfect way to collect potentially customer data— like emails— while also allowing them to inform even more potential customers about your brand.

Charity

Charity events are an excellent place for exposure of the small business brand — and for a good reason! Potential customers would equate your company with a good feeling and you will only have to donate time or some stuff.

Join a professional organization

A trade chamber or other B2B association is a great free marketing idea to create camaraderie for all involved parties. Engaging with other companies would give your company more exposure and provide incentives for engaging with other businesses in your region.

Mascot

Chester Cheeto, Ronald McDonald, Geico Lizard — the mascots expand the consumer base's brand awareness and make customers feel as though they are dealing with a product rather than being a faceless company. Consider creating your own mascot brand to promote your business better.

Controversial

Politics and business are connected than ever, and it is a great way to get their focus to stand on an issue that is essential to your consumer base.

Anti-tobacco? Stop selling cigarettes and publish a press release in your store.

Environmental care? Take the steps to make the company environmentally friendly, and then write a post about the process on a website such as Medium.

Take a position, act on it, and then talk it to people. While you may lose certain clients, you may improve and develop new partnerships with other customers.

Yelp

This is a basic yet powerful marketing idea for small companies managing brick and mortar stores or service companies. Yelp is a great way to find new customers as long as you remain reliable.

Yelp for company can be a paying or free marketing idea based on how much you want to spend. A first basic step is either to register or claim your company, add your location, post pictures of your products and begin to respond to customer reviews.

Forums

If your clients are online, the small business' best marketing plans will also be online.

Establish a profile on online forums such as Quora, Twitter, or almost every site that your clients share.

Answering questions in a non-promotional yet positive way will help to render the potential customers noticeable. Nonetheless, marketing a company so aggressively will drive potential new customers away.

Vehicle branding

When you run a business that involves driving around town all the time, branding your car will be an incredibly simple, almost free marketing idea for your small business.

With your logo on your car, you get name recognition in your local area and it's like free advertising after the initial investment.

Have a vehicle fleet? You should buy bulk car magnets and position them on all the vehicles.

Awards

There are some industry competitions you may qualify for, and if you receive, they can help get the reputation of your organization in the media. Often local newspapers praise companies or create a list of popular companies— another great way to get your company out. Search for the new honors or classifications in your field, so you don't miss a chance to make your company shine.

Awards such as the best places to work in your region or your industry are also a great way to make your business known. This goes a long way for potential customers to prove that you handle your workers fairly — consumers like to help companies that they invest in.

CONCLUSION

Now that we're done, I very much hope that this book has served you.

REMEMBER. You have studied hard to do your job, you know that nothing is given in life, work hard and apply the principles of this book, the results will come.

I wish you the best, and thanks again.

About the autor

C.J. Locke is the pen name of a personal brand marketing expert.
C.J. he started by washing floors in the offices of large advertising agencies.
He is now the owner of one of the largest advertising agencies in the world.
If he did it, anyone can do it.Love Space Invader, sneakers and ancient coins collection.